TABLE OF CONTENTS

III
IV
01
02
03
04

I

III. INTRODUCTION

"Believe in yourself and good things do start happening."
- David J Schwartz

HELLO!

As a former secondary mathematics teacher, I am thrilled to introduce my latest creation - a travel-sized interactive algebra workbook. This compact yet comprehensive guide is specifically designed to make learning algebra engaging and accessible for students on the go. Packed with interactive elements, it provides a hands-on approach to understanding key algebraic concepts, from basic equations to more complex functions. The workbook is structured to cater to a variety of learning styles, ensuring that each student can find a method that resonates with them. Its contents are carefully curated based on my years of experience in the classroom, ensuring that the material is not only educational but also aligns with current academic standards.

This workbook is more than just a learning tool; it's a portable companion for any student embarking on the exciting journey of algebra.

HOW TO USE THIS WORKBOOK

To maximize your learning with this interactive algebra workbook, simply scan the QR codes at the back of the book to access step-by-step video solutions for the problems. Whether you're stuck on a tricky equation or just want to reinforce your learning, these QR-linked videos are a handy resource for on-the-spot guidance.

IV. PREREQUISITE SKILLS
Assessment

Solve and record your answers. After completing, check your answers against the solutions in the back of this book. Good luck!

1. Simplify. $$2^4$$	**2. Simplify.** $$\sqrt{25}$$
3. Simplify. $\sqrt[3]{64}$	**4. Solve** $3x - 7 = 14$.
5. Solve. $$-a + 7 = 2a - 8$$	**6. Solve.** $$7(x - 2) = 7x + 14$$
7. Factor. $x^2 - 36$	**8. Solve** $y^2 + 6y + 8 = 0$.

9. Evaluate the function for x=3. $f(x) = -2x^2 + x - 5$	**10. Simplify.** $\dfrac{5x + 10}{2x + 4}$
11. Evaluate. $3^{x+1} = 81$	**12. Simplify.** $(2x^3y)^6$
13. Simplify. $(2x^2 + 3x - 1) + (3x^2 - 2x + 4)$	**14. Solve and graph.** $-3(x + 1) > 15$ 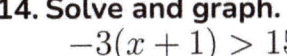
15. Graph. $y = 2x + 1$ 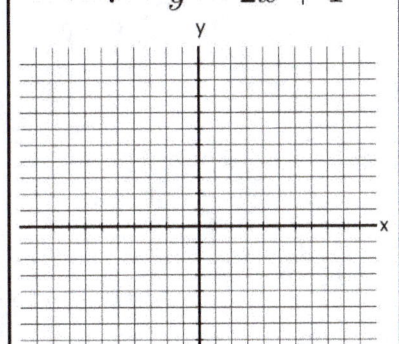	**16. Solve.** $\begin{cases} 2x + 3y = 9 \\ x - y = 2 \end{cases}$

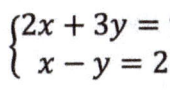

01. FOUNDATIONS OF ALGEBRA

1. Write an algebraic expression. "The difference of x and 5."	2. Solve the equation. $$y - 8 = 24$$
3. Solve. $$5q - 13 = 37$$	4. Solve. $$\frac{x+8}{-3} = -2$$
5. Solve. $$-3c - 12 = -5 + c$$	6. Solve. $$4x + 2y = 6, \text{ for } y$$

7. Solve. $\dfrac{21}{2x} = \dfrac{7}{10}$	8. Write in slope intercept form. $2x + y = 5$

9. Identify the intercepts.

$$3x + 5y = 15$$

10. Graph.

$$y = \frac{1}{5}x - 3$$

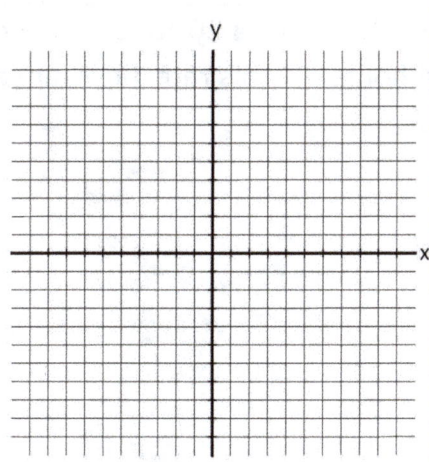

11. Write in slope-intercept form an equation of the line through (1, –3) and parallel to y = -6x – 2.

---

12. State the domain.

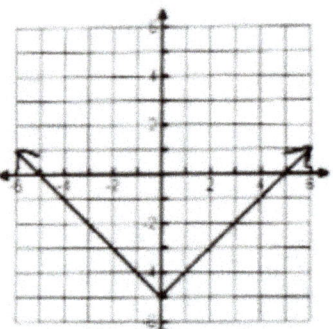

13. Solve.
$$|x + 3| = 4$$

---

14. Jeff rides a 20-mile trail every Saturday. It takes him 4 hours. At this rate, how far can he ride in 7 hours?

01. FOUNDATIONS OF ALGEBRA

Level 2

1. Write an algebraic expression. "The product of 9 and the sum of a number and 5."	**2. Solve the equation.** $$-\frac{2}{5} + x = -\frac{3}{5}$$
3. Solve. $$7p + 5 = 52$$	**4. Solve.** $$2f + 3(f + 4) = 32$$
5. Solve. $$20 - 8(g - 5) = 3g + 2(g + 4)$$	**6. Solve for g.** $$\frac{2 + x}{g} = 3y$$

7. Solve. $\dfrac{2h+5}{2} = \dfrac{4-h}{10}$	8. Write in slope intercept form. $-3x + 4y = -15$

9. Identify the intercepts.

$$5x - 2y = 10$$

10. Graph. $4x + 2y = -6$

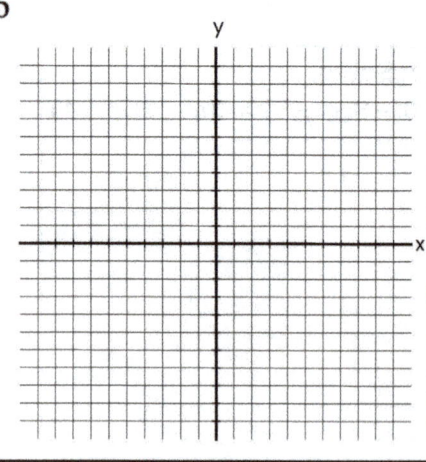

11. Write in slope-intercept form an equation of the line through (8, 5) and perpendicular to $y = -\dfrac{1}{4}x + 6$.

12. State the domain.

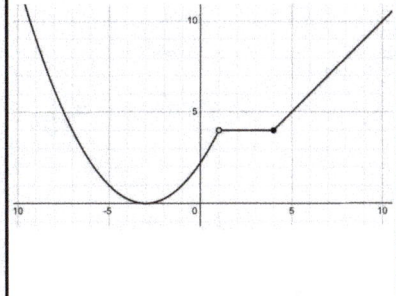

13. Solve.

$$2|x - 5| - 4 = 12$$

14. What is the height of a cylindrical paint bucket that has a radius of 6 inches and a volume of 522 cubic inches? Answer in terms of pi. (Note: $V = \pi r^2 h$)

01. FOUNDATIONS OF ALGEBRA

1. Write an algebraic equation and solve. The quotient of a number and 3 increased by 10 equals the sum of the same number and 2.	2. Solve the equation. $$\frac{5}{8}w = -20$$
3. Solve. $$\frac{t}{12} - 9 = -11$$	4. Solve. $$4 - 2(9x + 3) = 16$$
5. Solve. $$8x + 2 = 3(x - 5) - 7x$$	6. Solve for w. $$V = \frac{1}{3}whl$$

7. Solve. $\dfrac{x-1}{2x+3} = \dfrac{12}{11}$	8. Write in slope intercept form. $y+1 = 3(x-2)$

9. Identify the intercepts. $-3x + 5y = 30$

10. Graph both equations.

$x = -2$

$y = 3$

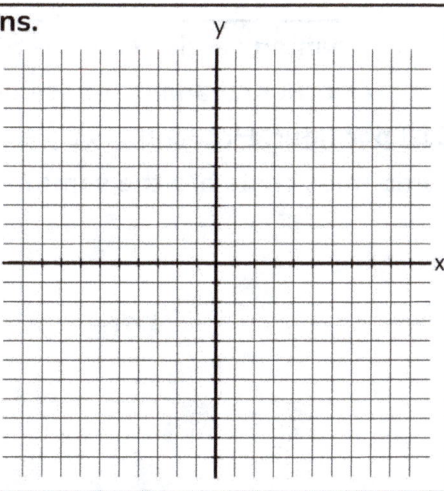

11. Determine if the lines are parallel, perpendicular, or neither.

$$3x + 4y = 12$$
$$8x - 6y = -60$$

12. State the range.

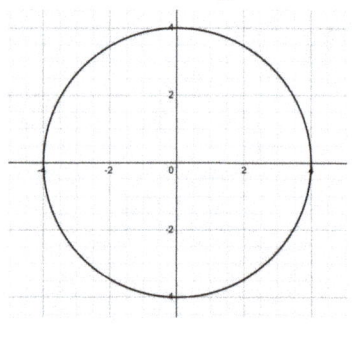

13. Solve.

$$\frac{1}{2}\left|\frac{2}{3}(x-4)\right| + 5 = 9$$

14. A cleaning service charges $150 base plus $16 per hour. Another cleaning service charges $110 base plus $18 per hour. How long is a job for which the two companies costs are the same?

02. INEQUALITIES

1. Write an inequality for the expression. "The product of 12 and a number is less than 6".	**2. Graph.** $x \geq 3$ ←—+—+—+—+—+—+—+—+—+—+—+—→
3. Write the inequality shown by the graph. 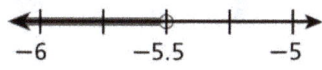 $-6 \quad -5.5 \quad -5$	**4. Solve and graph the solution.** $$-4 + x \leq -5$$ ←—+—+—+—+—+—+—+—+—+—+—+—→
5. Solve. $2m \geq 6$	**6. Solve.** $3x - 8 > -1$

15

7. Write the inequality and solve.	8. Solve and graph.
"Three more than half a number is less than fifteen."	$z + 3 \geq 1$ and $z - 2 < -1$ 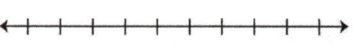

9. Solve the inequality.

$$|2x - 1| > 5$$

10. You have been saving for a new wallet that costs no less than $200. You have saved $130. Write and solve an inequality to determine that amount you need to purchase the phone.

02. INEQUALITIES

1. Write an inequality for the expression. "The sum of a number and 2 is no less than the product of 9 and the same number."	2. Graph. $$4 \geq x$$
3. Write the inequality shown by the graph.	4. Solve and graph the solution. $$d + \frac{4}{7} > \frac{1}{14}$$
5. Solve. $$\frac{1}{3}x < \frac{2}{3}$$	6. Solve. $$\frac{x+3}{-5} > 2$$

7. Solve. $-2(3x + 1) > -6x + 7$	8. Solve and graph. $3 < 5x - 2 < 13$

9. Solve the inequality.

$$|4x + 7| \leq 19$$

10. You work at a local car wash and make $8.50 per hour. How many hours do you need to work to make at least $150?

02. INEQUALITIES

1. Write an inequality for the expression. "The quotient of a number and 5 is at least 10."	2. Graph. $x < -6 \ or \ x \geq 2$
3. Write the inequality shown by the graph. -10 ———○———-5———————○——— 0	4. Solve and graph the solution. $w + 2.8 \leq 13.9$
5. Solve. $-63n < 7$	6. Solve. $-4(2 - x) \leq 8$

7. Solve.	8. Solve and graph.
$5(2x - 3) - 7x \leq 3x + 8$	$p + 5 \geq 10$ or $-2p > 10$

9. Solve the inequality. $8\lvert x + 4 \rvert + 10 < 2$

10. For the weekend, you and your friends go to a theme park. Admission to the park costs $32.50 and each ride costs $0.80. You have $50 to spend at the park including admission. What are the maximum number of rides you can enjoy?

03. FUNCTIONS AND RELATIONS

1. Determine the domain and range for each relation. $\{(2, 3), (-1,5), (-5, 5), (0, -7)\}$	2. Determine whether the relation is a function. $\{(0, 1), (1, 0), (2, 1), (3, 1), (4, 2)\}$
3. Evaluate $f(x) = -2x + 11$ for f(4), f(-5), and [8 – f(0)].	4. Find $h(x) = f(x) + g(x)$. Given: $f(x) = 3x - 1$ $g(x) = -2x + 2$
5. Find $h(x) = f(x) - g(x)$. Given: $f(x) = 16x - 2$ $g(x) = -5x - 3$	6. Find $h(x) = f(x) \cdot g(x)$ Given: $f(x) = 3$ $g(x) = \frac{1}{3}x - 2$

| 7. Find $f(x) = g(x)$
Given: $f(x) = 4x - 6$
$g(x) = 2x + 2$ | 8. Evaluate the piecewise function.
$f(x) = \begin{cases} 4x + 4 & \text{if } x < 0 \\ x + 7 & \text{if } x \geq 0 \end{cases}$

a) $f(-2) = $ _____
b) $f(0) = $ _____
c) $f(2) = $ _____ |

9. Graph.

$f(x) = \begin{cases} x + 3 & \text{if } x \leq 0 \\ 2x & \text{if } x > 0 \end{cases}$

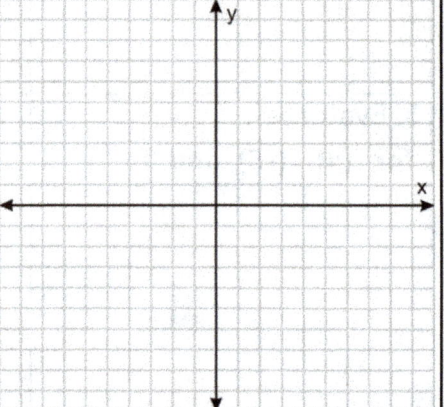

10. Describe the end behavior of the graph.

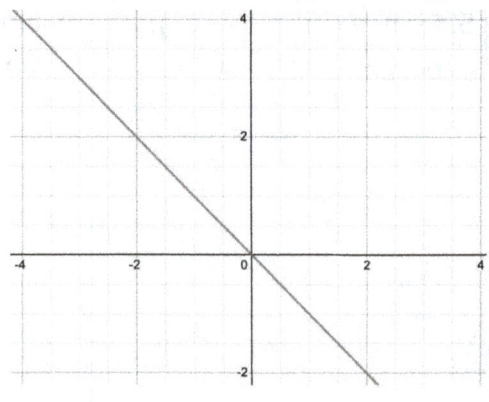

03. FUNCTIONS AND RELATIONS

1. Determine the domain and range for the relation.	2. Determine whether the relation is a function.
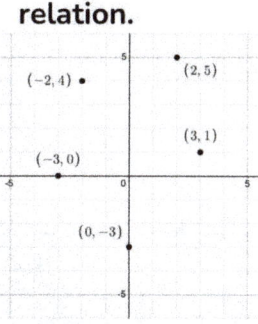	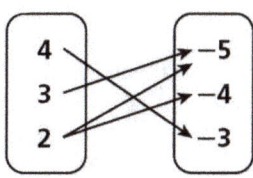
3. Evaluate $f(x) = x^2 + 5x - 6$ for f(2), f(-1), and [f(0) + f(1)].	4. Find $h(x) = g(x) + f(x)$ Given: $f(x) = 4x - 2$ $g(x) = x + 5$
5. Find $h(x) = f(x) - g(x)$. Given: $f(x) = 4x + 6$ $g(x) = 12x - 8$	6. Find $h(x) = g(x) \cdot f(x)$. Given: $f(x) = -2$ $g(x) = \frac{3}{4}x - 5$

7. Find $f(x) = g(x)$ Given: $f(x) = -3x + 3$ $g(x) = 2x - 7$	8. Evaluate the piecewise function. $$f(x) = \begin{cases} x+3 & \text{if } x < 0 \\ -2x & \text{if } x \geq 0 \end{cases}$$ $f(-1), f(0.5), f(7), \text{ and } f(0)$

9. Graph.
$$f(x) = \begin{cases} 2x+3 & \text{if } x < 4 \\ x-1 & \text{if } x \geq 4 \end{cases}$$

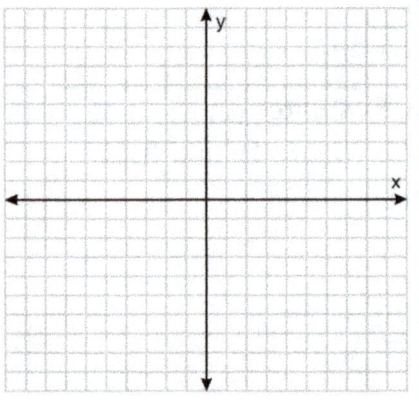

10. Describe the end behavior of the graph.

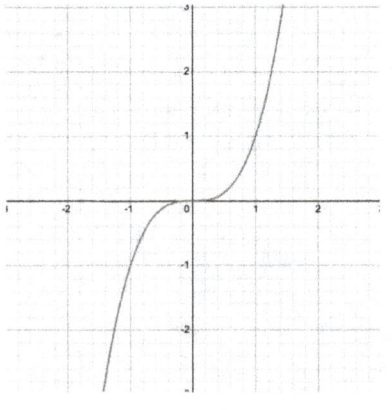

03. FUNCTIONS AND RELATIONS

1. Determine the domain and range for each relation. 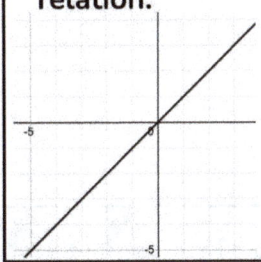	2. Determine whether the relation is a function. 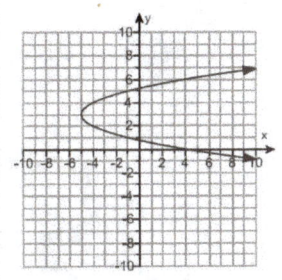
3. Find the range of $f(x) = (x-1)^2$ when the domain is {-3,0,5}.	4. Find $h(x) = f(x) + g(x)$. Given: $f(x) = 3x - 1$ $g(x) = 6x + 2$
5. Find $h(x) = f(x) - g(x)$. Given: $f(x) = -x + 3$ $g(x) = 2x + 6$	6. Find $h(x) = f(x) \cdot g(x)$ Given: $f(x) = -4$ $g(x) = -\frac{1}{2}x + 2$

7. Find $f(x) = g(x)$	8. Evaluate the piecewise
Given: $f(x) = 4x - 10$ $g(x) = 3x + 6$	**function.** $f(x) = \begin{cases} -x & \text{if } x < -1 \\ 2x + 1 & \text{if } -1 \le x < 0 \\ x^2 & \text{if } x \ge 0 \end{cases}$ $f(-3), f(0.2), f(5), and\ f(10)$

9. Graph.

$$f(x) = \begin{cases} -3 & \text{if } x < 0 \\ -\dfrac{1}{2}x + 2 & \text{if } 0 \le x \le 2 \\ 4 & \text{if } x > 2 \end{cases}$$

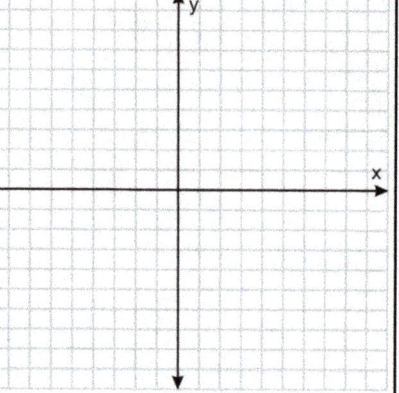

10. Describe the end behavior of the graph.

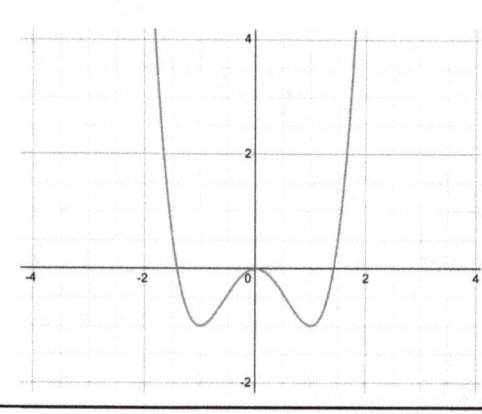

04. SYSTEMS OF EQUATIONS

1. Graph and determine the number of solutions.
$$\begin{cases} y = -4x + 2 \\ y = x - 2 \end{cases}$$

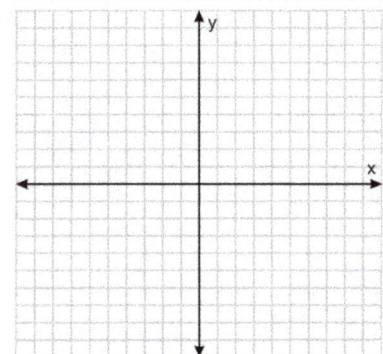

2. Solve the system by substitution.
$$\begin{cases} -2x + y = -2 \\ x = y + 1 \end{cases}$$

3. Solve the system by elimination.
$$\begin{cases} -9x + y = 3 \\ 9x - 3y = 17 \end{cases}$$

4. Determine the best method for the system and solve.

$$\begin{cases} 3x + y = -16 \\ -3x + 4y = 1 \end{cases}$$

5. Graph.

$y \leq 3x - 1$

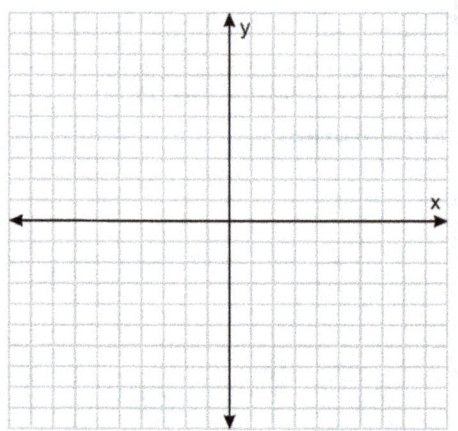

6. Graph.

$$\begin{cases} -x + y < -1 \\ x + y < 3 \end{cases}$$

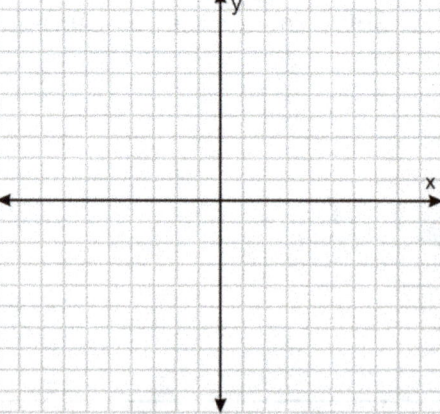

7. A local coffee shop sold a total of 35 tickets for an open mic event. Early admission tickets were $5 each and general admission tickets were $8 each. If the coffee shop made a total of $238, how many of each ticket did they sell?

8. Jeff has $44 to buy 7 books. Paperback books cost $5 each. Hardcover books cost $8 each. How many of each book can Jeff buy?

04. SYSTEMS OF EQUATIONS

1. Graph and determine the number of solutions.

$$\begin{cases} x+y=4 \\ y=-x+1 \end{cases}$$

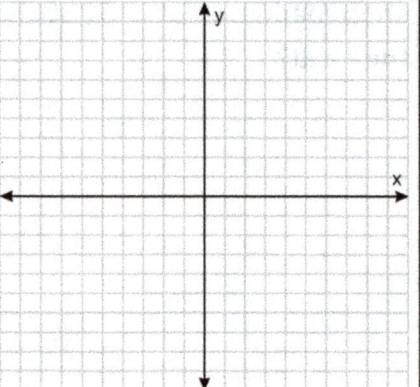

2. Solve the system by substitution.

$$\begin{cases} -y=-2x+4 \\ -6x+3y=-12 \end{cases}$$

3. Solve the system by elimination.

$$\begin{cases} x+3y=12 \\ 4x+y=-7 \end{cases}$$

4. Determine the best method for the system and solve.

$$\begin{cases} x = -4y + 5 \\ 3x - 7y = 8 \end{cases}$$

5. Graph $y > 4$.

   Is (2,4) a solution?

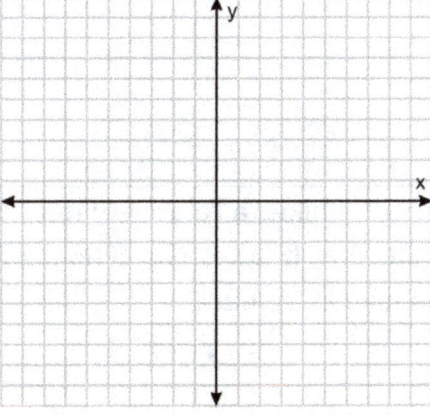

6. Graph and give a solution.

$$\begin{cases} -2x + y < -1 \\ y > 2x + 3 \end{cases}$$

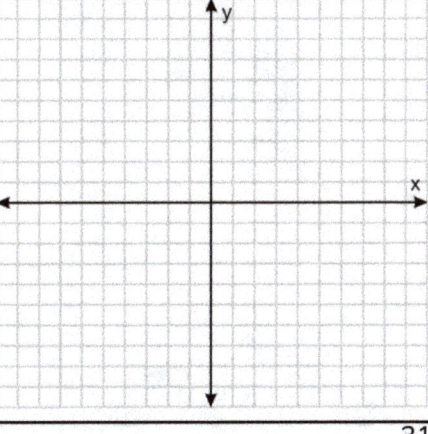

7. The sum of two angles is 180°. The measure of one angle is 24° greater than the measure of the other angle. Define the variables, write equations, and find the measure of each angle.

8. Colin hikes a relatively flat 15 mile trail every Saturday. It takes him 5 hours. At this rate, how far can he hike in 7 hours?

04. SYSTEMS OF EQUATIONS

1. Graph and determine the number of solutions.

$$\begin{cases} y = 3x + 7 \\ -2y + 6x = -14 \end{cases}$$

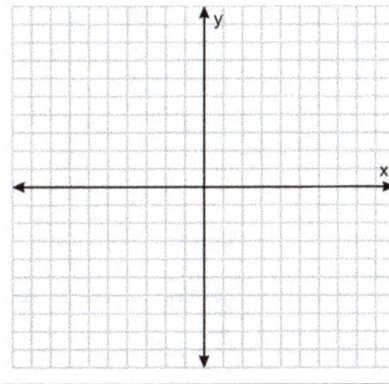

2. Solve the system by substitution.

$$\begin{cases} x + 3y = 5 \\ -2x + 4y = 0 \end{cases}$$

3. Solve the system by elimination.

$$\begin{cases} 2m + 4n = -4 \\ 3m + 5n = -3 \end{cases}$$

4. Determine the best method for the system and solve.
$$\begin{cases} 5x - 3y = 4 \\ 15x - 9y = 12 \end{cases}$$

5. Graph.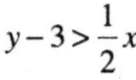

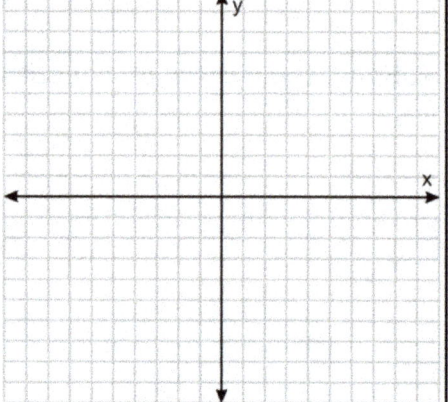

6. Graph.
$$\begin{cases} y \geq 3x - 4 \\ y \leq 3 \\ x > -2 \end{cases}$$

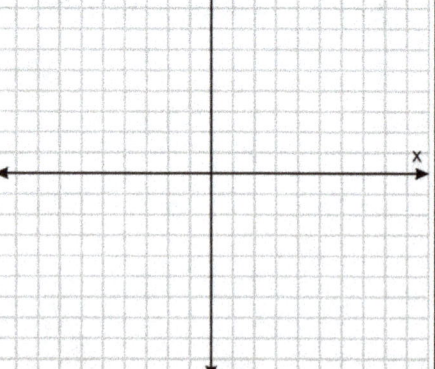

7. You have a coin jar that contains only dimes and quarters. The total value of these coins is $3.60. If you notice that the number of dimes is double the number of quarters, how many of each type of coin do you have?

8. You are mixing apple and orange juice for a party. You want to mix a total of 15 liters of juice. The amount of apple juice should be 2/3 the amount of orange juice. How many liters of each type of juice do you need?

05. QUADRATIC FUNCTIONS AND EQUATIONS

1. Write in standard form. Identify the leading coefficient and degree. $$4a^2 - 7a + 3a^5$$	2. Simplify. $(2x^2 - 7 + 5x) + (-4x^2 + 6x + 3)$
3. Simplify. $$(4h + 5j) - 3h$$	4. Simplify. $$2x^2(6x^2 - 2x + 5)$$
5. Simplify. $(x + 3)(x + 9)$	6. Simplify. $$(2x - 6)(3x^2 + x - 1)$$

7. Simplify. $(x+2)^2$	8. Factor. $15x^2 + 100$
9. Factor. $x^3 + 2x^2 - 3x - 6$	10. Factor. $x^2 + 10x + 16$
11. Factor. $x^2 + 10x + 25$	12. Solve. $2x^2 = 128$

13. Solve. $\sqrt{25}$	14. Solve. $5x^2 = 20$

15. Solve by completing the square. $x^2 + 6x - 27 = 0$

16. Find the zeros using the quadratic formula.

$$x^2 + 4x - 32 = 0 \qquad x = \frac{-b \pm \sqrt{b^2 - 4ac}}{2a}$$

05. QUADRATIC FUNCTIONS AND EQUATIONS

1. Write in standard form. Identify the leading coefficient and degree. $5h - 9 - 2h^4 - 6h^3$	2. Simplify. $(5x + 7x^2 + 3) + (-5x^2 + x^3 - 4)$
3. Simplify. $(3x + 2 - x^2) - (4x - 5 + 2x^2)$	4. Simplify. $3(5x^2 + x - 4) - x(4x^2 + 2x - 3)$
5. Simplify. $(5w - 2)(w + 3)$	6. Simplify. $(b^2 - 4b + 3)(b - 2)$

7. Simplify. $(x^2+6)(x^2-6)$	8. Factor. $6x^3 - 9x^2 + 12x$
9. Factor. $x^3 + 4x + x^2 + 4$	10. Factor. $5x^2 - 13x + 6$
11. Factor. $x^2 - 100$	12. Solve. $2x^2 + 50 = 20x$

13. Solve. $\sqrt{72}$	14. Solve. $4x^2 - 25 = 0$

15. Solve by completing the square.

$$3x^2 - 24x - 27 = 0$$

16. Find the zeros using the quadratic formula.

$$2x = x^2 - 3 \qquad x = \frac{-b \pm \sqrt{b^2 - 4ac}}{2a}$$

05. QUADRATIC FUNCTIONS AND EQUATIONS

1. Write in standard form. Identify the leading coefficient and degree. $-5 + 7x^2 - 14x + 6x^2$	2. Express the perimeter as a polynomial. $x^2 - 9$ $4x^2 - 3x + 8$ $3x^2$
3. Simplify. $(12x^2 - 8x + 11) - (-14 + 10x^2 - 6x)$	4. Express the area as a polynomial. $2x - 5$ $4x^2 + x + 3$
5. Simplify. $(4k-1)(3k-7)$	6. Simplify. $(x+2)[(x^2 + 3x - 6) + (x^2 - 2x + 4)]$

7. Simplify. $\left(\dfrac{1}{4}x+2\right)^2$	8. Factor. $14x^3y + 7x^2y - 7xy$
9. Factor. $xy - 6x + 6y - 36$	10. Factor. $-2x^2 - 5x - 3$
11. Factor. $x^2 + 4$	12. Solve. $21x^2 = 3x^3 + 36x$

13. Solve. $\sqrt{80}$	14. Solve. $8(x-1)^2 - 5 = 27$

15. Solve by completing the square. $4x^2 - 32 = 24x$

16. Evaluate the discriminant and number of real solutions.

$3x^2 - 7x + 1 = 0$

If $b^2 - 4ac > 0$, the equation has 2 real solution(s).
If $b^2 - 4ac = 0$, the equation has 1 real solution(s).
If $b^2 - 4ac < 0$, the equation has 0 real solution(s).

06. QUADRATIC FUNCTIONS AND GRAPHS

1. Use the table of values to graph the parabola.

$f(x) = x^2$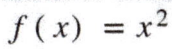

x	y
2	
1	
0	
-1	
-2	

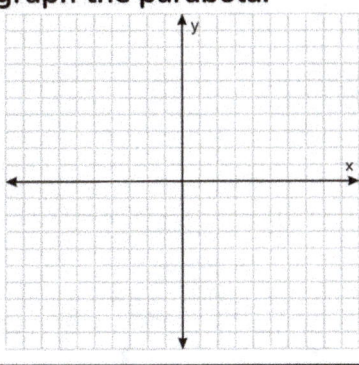

2. Complete the table.

Vertex	
Min/Max	
Axis of symmetry	
x-intercept(s)	
y-intercept	

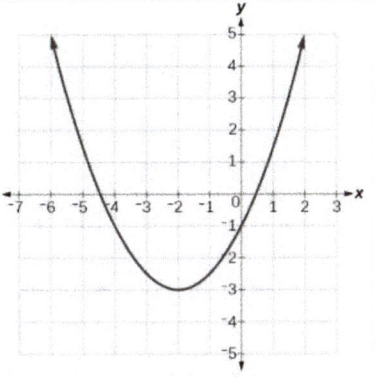

3. Find the vertex, AOS, and y-intercept. Then, graph.

$f(x) = x^2 - 6x + 5$

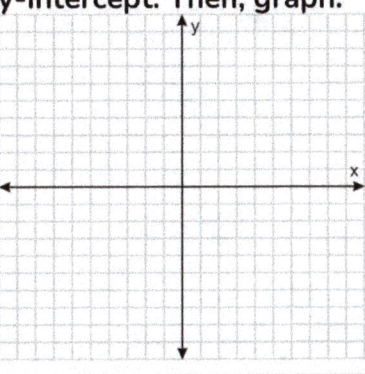

4. Change to Vertex Form. Identify the vertex, AOS, and y-intercept. $f(x) = x^2 + 2x - 3$

> Vertex form:
> $f(x) = a(x - h)^2 + k$

5. Find the vertex, AOS, y-int, range, and graph.

$f(x) = (x + 2)^2 - 5$

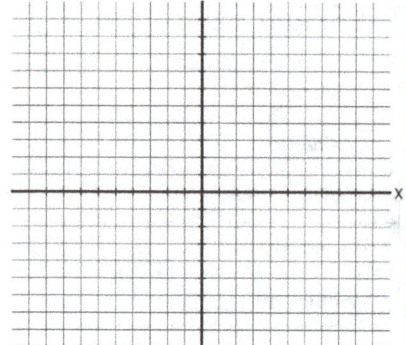

6. Write a quadratic function in vertex form to model the graph.

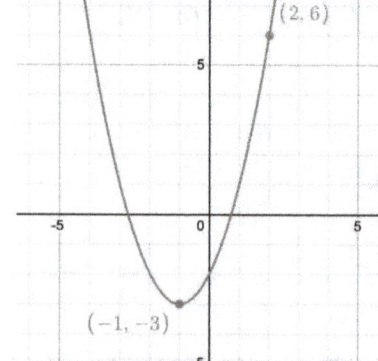

$(2, 6)$

$(-1, -3)$

7. Solve the system by graphing.

$$\begin{cases} y = x^2 - 2x + 4 \\ 2x + y = 5 \end{cases}$$

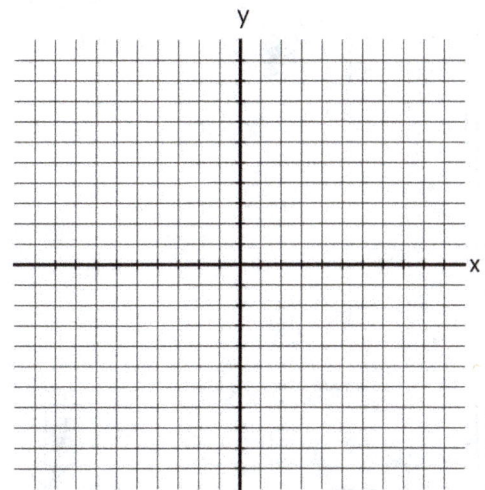

8. A ball is thrown straight up into the air from the top of a building. The height of the ball above the ground, in feet, after t seconds is given by the quadratic function $h(t) = -16t^2 + 64t + 80.$ Determine how long it takes for the ball to hit the ground.

06. QUADRATIC FUNCTIONS AND GRAPHS

1. Use the table of values to graph the parabola.

x	y
0	
1	
2	
3	
4	

$f(x) = x^2 - 4x + 1$

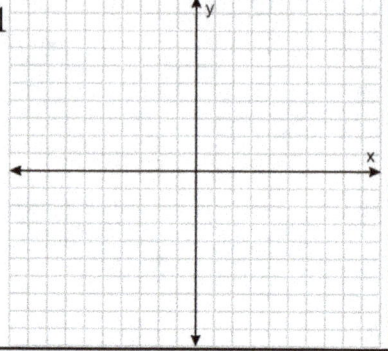

2. Complete the table.

Vertex	
Max/Min	
AOS	
x-intercept(s)	
y-intercept	

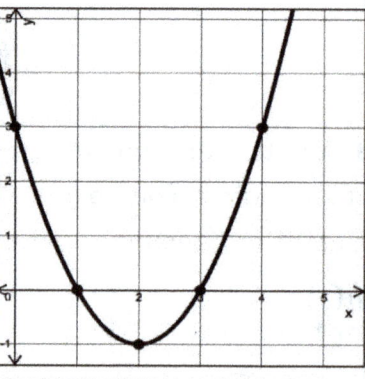

3. Find the vertex, AOS, and y-intercept. Then, graph.

$f(x) = 2x^2 + 8x + 7$

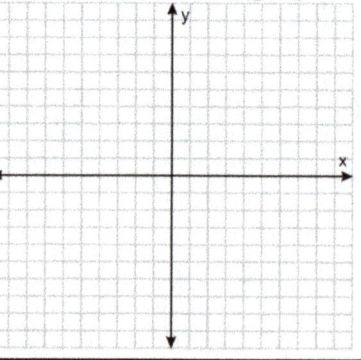

4. Change to Vertex Form. Identify the vertex, AOS, and y-intercept. $f(x) = x^2 + 12x + 4$

Vertex form:
$f(x) = a(x - h)^2 + k$

5. Find the vertex, AOS, y-int, range, and graph.

$f(x) = -(x - 1)^2 + 4$

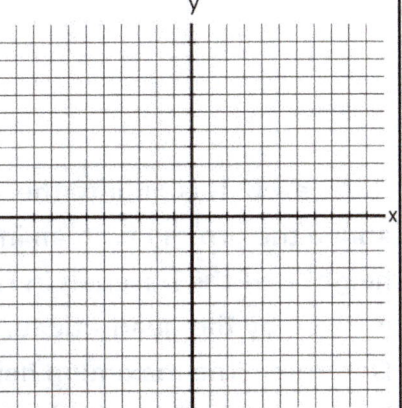

6. Write a quadratic function in vertex form to model the graph.

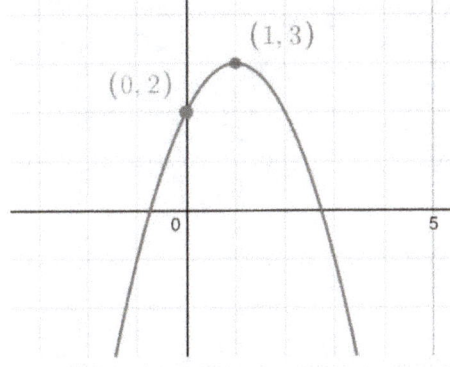

$(1, 3)$

$(0, 2)$

7. Solve the system algebraically.

$$\begin{cases} y = -x^2 + 4x \\ \quad y = -5 \end{cases}$$

8. A baseball is hit into the air from a height of 3 feet above the ground with an initial upward velocity. The height h of the baseball in feet, at any time t in seconds, is given by the quadratic function $h(t) = -16t^2 + 32t + 3$. Determine the maximum height of the baseball, the time it takes to reach its maximum height, and the total time it is in the air before it hits the ground.

06. QUADRATIC FUNCTIONS AND GRAPHS

1. Use the table of values to graph the parabola.

x	y
-3	
-2	
-1	
0	
1	

$f(x) = x^2 + 2x + 3$

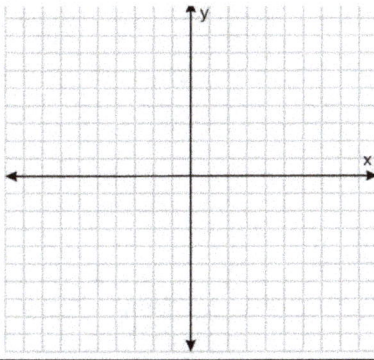

2. Complete the table.

Vertex	
Max/Min	
Axis of symmetry	
x-intercept(s)	
y-intercept	

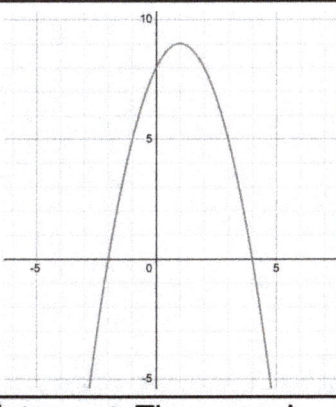

3. Find the vertex, AOS, and y-intercept. Then, graph.

$f(x) = -2x^2 + 4x - 2$

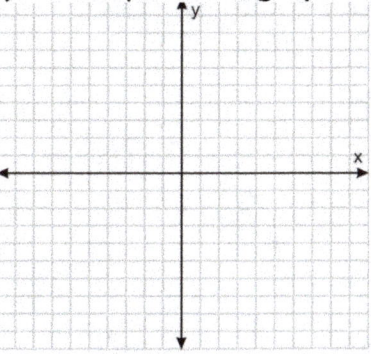

4. Change to Vertex Form. Identify the vertex, AOS, and y-intercept. $f(x) = -x^2 - 6x + 2$

Vertex form:
$f(x) = a(x - h)^2 + k$

5. Find the vertex, AOS, y-int, range, and graph.
$f(x) = 2(x + 1)^2 + 3$

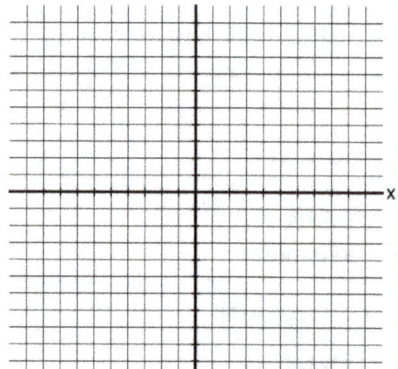

6. Write a quadratic function in vertex form to model the graph.

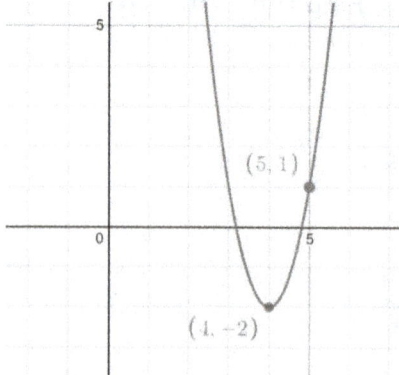

$(5, 1)$

$(4, -2)$

7. **Solve the system by graphing or algebraically.**

$$\begin{cases} y = x^2 - 4x + 1 \\ \quad y = -4 \end{cases}$$

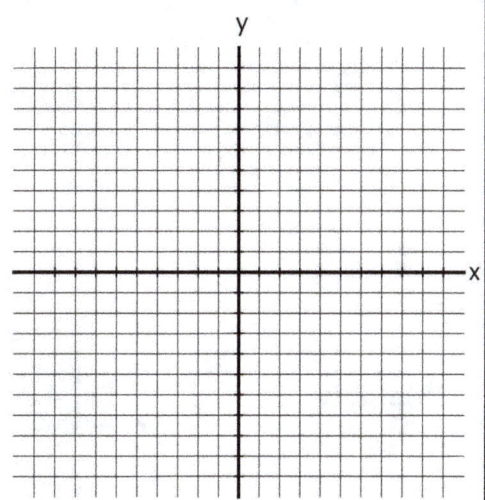

8. **Complete the blank spaces below to describe the transformations from the parent function f(x).**

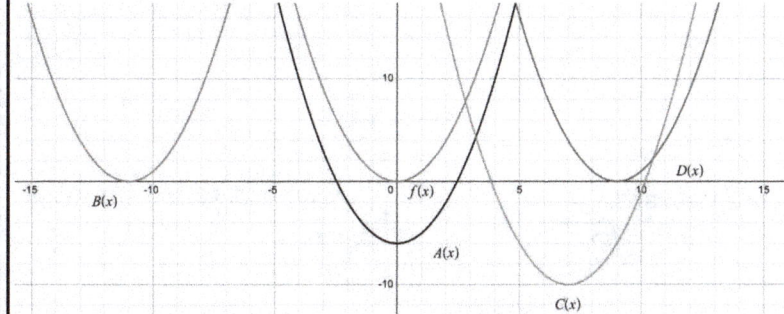

$A(x) = (x - \underline{\hspace{1cm}})^2 + \underline{\hspace{1cm}}$

$B(x) = (x - \underline{\hspace{1cm}})^2 + \underline{\hspace{1cm}}$

$C(x) = (x - \underline{\hspace{1cm}})^2 + \underline{\hspace{1cm}}$

$D(x) = (x - \underline{\hspace{1cm}})^2 + \underline{\hspace{1cm}}$

07. EXPONENTS AND RADICALS

1. Evaluate $x^2y^3 + 2x$ if $x = -1$ and $y = 10$.	2. Simplify. 3^{-3}
3. Simplify. $(-2)^{-2}$	4. Simplify. $8x^{-4}$
5. Evaluate. $\sqrt{36}$	6. Convert to exponential form. $\sqrt[4]{x}$

7. Convert to radical form. $x^{\frac{3}{7}}$	8. Evaluate. $8^{\frac{2}{3}}$
9. Solve. $12^x = 144$	10. Simplify. $\sqrt{x^2 y^6}$
11. Simplify. $\sqrt{20}$	12. Simplify. $\sqrt{x^5}$

13. Simplify. $$\frac{y^{11}}{y^2}$$	14. Simplify. $$x^{\frac{2}{5}} \cdot x^{\frac{1}{10}}$$
15. Simplify. $$16^{-\frac{1}{2}}$$	16. Simplify. $$\sqrt{72xy} \cdot \sqrt{4xy^2}$$

17. Find the inverse.
$$f(x) = 4x + 1$$

07. EXPONENTS AND RADICALS

1. Evaluate $\dfrac{3x^2}{4y^2}$ if $x = 4$ and $y = (-7)$.	**2.** Simplify. 10^0
3. Simplify $-6a^0b^{-2}$ if $a=5$ and b= -3.	**4.** Simplify. $\dfrac{-5}{k^{-2}}$
5. Evaluate. $\sqrt[3]{8}$	**6.** Convert to exponential form. $\sqrt{a^5}$

7. Convert to radical form. $$3x^{\frac{3}{2}}$$	8. Evaluate. $$\left(\frac{1}{81}\right)^{\frac{1}{4}}$$
9. Solve. $$3^x = 27$$	10. Simplify. $$\sqrt[3]{a^9 b^{15}}$$
11. Simplify. $$\sqrt[3]{32}$$	12. Simplify $$\sqrt[3]{x^{17}}$$

13. Simplify.

$$\frac{\sqrt[4]{y^6}}{\sqrt[5]{y^6}}$$

14. Simplify.

$$\sqrt{11} \cdot \sqrt[4]{11}$$

15. Evaluate.

$$\left(3x^{\frac{1}{5}}\right)\left(2x^{\frac{1}{3}}\right)$$

16. Simplify.

$$\left(\frac{3a^{-5}b^8}{10a^9b^{-9}}\right)^0$$

17. Find $(f \circ g)(x)$ **for** $f(x) = 2x - 4$ **and** $g(x) = 2x^2 - 1$.

07. EXPONENTS AND RADICALS

1. Evaluate $\dfrac{x^{-2} \cdot y^{-3}}{x^{-5} \cdot y^{-1}}$ given that x = 2 and y = 3.	2. Simplify. $$-5^{-3}$$
3. Simplify if a = -2 and b = 6. $\quad 9a^{-3}b^0$	4. Simplify. $\dfrac{a^{-2}b^0}{c^{-6}d^3}$
5. Evaluate. $$\sqrt[5]{100,000}$$	6. Convert to exponential form. $\sqrt[3]{(2x)^7}$

7. Convert to radical form. $(4a)^{\frac{3}{5}}$	8. Evaluate. $625^{\frac{3}{4}}$
9. Factor. $3^x = 243$	10. Simplify. $\sqrt[3]{27y^{12}z^{14}}$
11. Simplify. $\sqrt[4]{48}$	12. Solve. $\sqrt[4]{324c^8d^6}$

13. Simplify. $\dfrac{25x^3y^4}{30x^6y^{-2}}$	14. Simplify. $\sqrt{\dfrac{9}{25}}$
15. Simplify. $(-216x^{-12})^{\frac{1}{3}}$	16. Rationalize the denominator. $\dfrac{\sqrt{2}}{\sqrt{3}}$

17. Find $(f \circ g)(-1)$ for $f(x) = 3x - 3$ and $g(x) = 4x^2 - 3$.

08. EXPONENTIAL FUNCTIONS

1. Write the next three terms of the arithmetic sequence. 1, 4, 7, 10,...	2. Determine if the data represents a linear or exponential function. Explain. <table><tr><td>x</td><td>1</td><td>2</td><td>3</td><td>4</td></tr><tr><td>y</td><td>3</td><td>6</td><td>12</td><td>24</td></tr></table>
3. Identify exponential growth or decay. Then find the y-intercept. $y = \dfrac{1}{3} \cdot (2)^x$	4. A small town currently has a population of 8,500 people. Calculate the population in 5 years if the annual growth rate is 2.5%.

5. The annual city chess championship is organized in a knockout format. The tournament begins with 256 players. In each round, half of the participants are eliminated. How many players are left after 4 rounds?

6. Sarah decides to invest $5,000 in a savings account that offers a 4% annual interest rate, compounded quarterly. How much money will Sarah have in the account at the end of 3 years?

7. Solve.
$$2^x = 8$$

8. Solve.
$$5^{2x+1} = 125$$

9. Solve.
$$8^{1-x} = 4^{x+2}$$

10. Solve.
$$2^{2y+2} = 16^{y-3}$$

08. EXPONENTIAL FUNCTIONS

1. Write the next three terms of the geometric sequence. 4, –8, 16, –32,...	2. Determine if the data represents a linear or exponential function. Explain. $y = 3x + 4$
3. Identify exponential growth or decay. Then find the y-intercept. $y = 0.455(3)^x$	4. A 500 grams radioactive substance decays at a rate of 6% per year. Determine how much of the substance will remain after 8 years.

5. A certain radioactive element has a half-life of 5 years. If you start with a 100-gram sample, how much of the substance will remain after 15 years?

6. A man deposits $30,000 in a savings account that earns an annual interest rate of 4%. Calculate the total interest earned and the amount in the account after 3 years, assuming the interest is compounded annually.

---

7. Solve.

$$3^x = 81$$

8. Solve.

$$4^{2-x} = \frac{1}{64}$$

---

9. Solve.

$$2^{x+2} \cdot 2^x = 16$$

10. Solve.

$$(7)^{3x} = 1$$

---

08. EXPONENTIAL FUNCTIONS

1. Write an equation for the nth term of the sequence. Then find the 8th term. −2, 7, 16,…	2. Determine if the data represents a linear or exponential function. Explain. 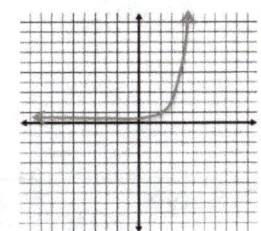
3. Identify exponential growth or decay. Then find the y-intercept. $y = 3 \cdot \left(\dfrac{2}{3} \right)^x$	4. A laptop initially costs $2,000 and depreciates at a rate of 20% per year. Calculate its value after 5 years.

5. You invest $1500 in a bond that offers a 3% annual interest rate, compounded semi-annually. What will be your balance after 4 years?

6. John obtained a car loan of $5,000 for a period of 3 years. At the end of the loan term, he had paid a total of $600 in interest. What was the annual interest rate on his loan?

7. Solve.
$$10^{x-6} = 1000$$

8. Solve.
$$6^{\frac{x-2}{5}} = \sqrt{6}$$

9. Solve.
$$3^{x^2-12} = 9^{2x}$$

10. If $\left(\frac{1}{3}\right)^{x-2} = 81^{x+1}$ find the value of $2x - 1$.

09. STATISTICS AND PROBABILITY

1. Find the mean of the following set of numbers: 5, 8, 12, 15, 18?	2. What is the median of the following set of numbers: 3, 7, 9, 12, 16?
3. What is the mode of the following set of numbers: 4, 4, 5, 7, 8, 8, 9?	4. What is the probability of drawing a red card from a standard deck of playing cards?

5. What is the probability of flipping a coin and getting heads and then rolling a die and getting a six?	**6. What is the range of the following set of numbers: 4, 7, 13, 17, 20?**

7. Use the histogram to answer the following:

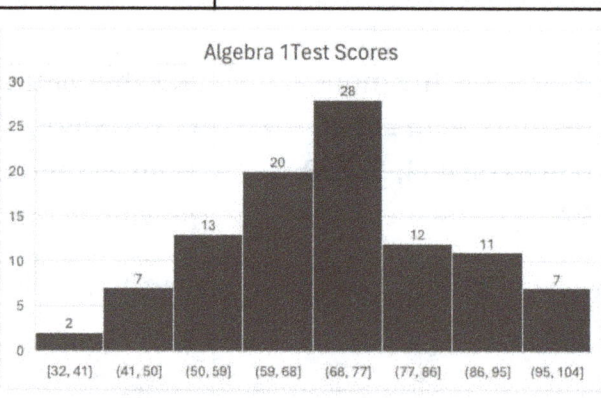

Algebra 1 Test Scores

a) How many test takers received a score between 50-59?	
b) How many people took the test?	
c) What percentage of test takers failed (less than 60)?	

8. Use the data representing the average daily temperatures (in degrees Fahrenheit) of a particular city over a month to create a box plot.

68	76	71	69	75	76	74	77
78	73	70	76	74	76	75	71

9. Sketch a normal curve for the distribution. Label the x-axis values at one, two and three standard deviations from the mean.   mean = 75,
standard deviation = 5

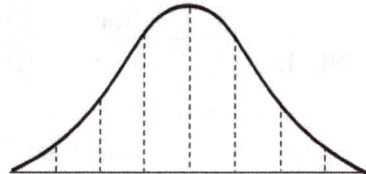

10. In a bag of 8 marbles, 3 are red, and 5 are blue. What is the probability of picking a red marble first and then a blue marble without replacement?

09. STATISTICS AND PROBABILITY

1. 40 11th and 12th grade members were surveyed about their favorite type of music for the upcoming school dance. Complete the frequency table.

	Favorite Type of Music				
Grade	Pop	Rock	Classical	Jazz	Total
11th	5	3		4	
12th	7		2		
Total		10		10	

2. Given the set of numbers: 32, 28, 35, 28, 40, and 32, find the mean, median, and mode.

3. Find the range and interquartile range for 160, 145, 135, 142, 155, and 140.

4. How many different ways can you arrange the letters in the word "STATS"?	5. Calculate the variance and standard deviation for 4, 8, 10, 12, and 20.

6. Create a histogram for the ages of CEOs of major tech companies at the time of their appointment.

| 54 | 42 | 51 | 56 | 55 | 54 | 51 | 60 | 62 | 43 |
| 56 | 61 | 52 | 69 | 64 | 54 | 47 | 51 | 55 | 46 |

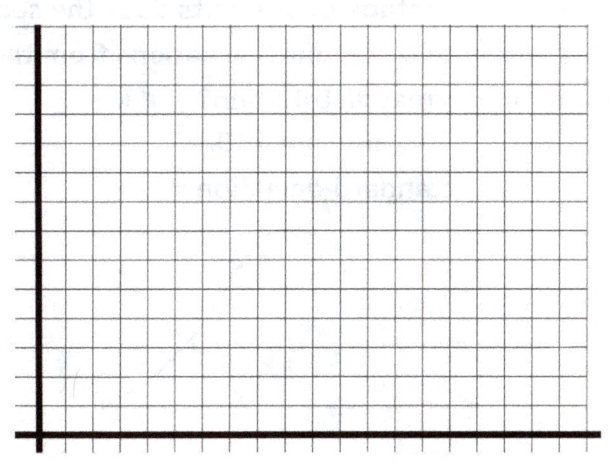

7. If a set of scores is normally distributed with a mean of 100 and a standard deviation of 15, what is the probability of a score being between 85 and 115?

8. Sketch the normal distribution curve for the heights of students in a biology class. Label the x-axis values at one, two and three standard deviations from the mean. What percentage of students does the section of students within two standard deviations from the mean represent in a normal distribution?

mean = 160,

standard deviation = 10

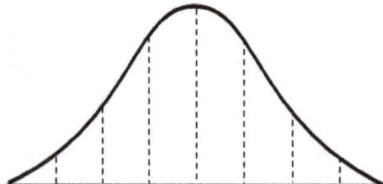

9. The table shows the relationship between the depth, d, of the lake in feet and the water temperature, t, in degrees Fahrenheit. Complete a-c.

d	t
0	85
4	82
7	81
10	79
12	77
15	76

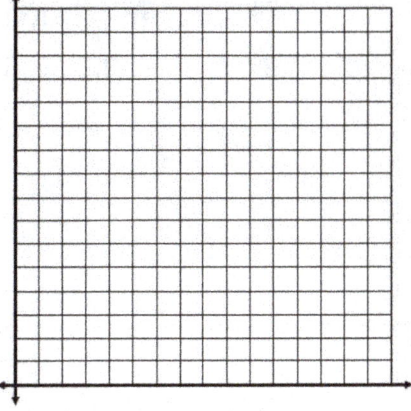

a) Make a scatter plot.

b) Use two ordered pairs to write a line of best fit.

c) Predict the temperature of the lake at 25 feet.

10. Gerardo packed 5 different t-shirts, 4 pants, and 3 hats. Assuming Gerardo chooses one of each type of clothing item for an outfit, how many different outfits can he create for his vacation?

09. STATISTICS AND PROBABILITY

1. Given a data set, calculate the mean, median, mode, and range.   3, 7, 9, 15, 21

2. A jar contains 3 red, 4 blue, and 5 green marbles. What is the probability of randomly drawing a green marble?

3. A bag contains 4 red balls and 8 black balls. Two balls are drawn randomly. What is the probability that both are red?

4. The recorded lengths from a sample of 15 fish are as follows: 27, 30, 24, 35, 28, 40, 22, 32, 29, 31, 25, 33, 26, 34, 50. Create a box plot to represent the distribution of fish lengths.

5. In a normal distribution with a mean of 30 and a standard deviation of 4, find the Z-score of a score of 34.

6. The table shows the number of hours 20 students spent studying for an Algebra exam. Create a histogram to represent this data.

Hours Studied	Number of Students
0-1	2
2-3	5
4-5	6
6-7	4
8-9	3

7. Evaluate. $$\frac{6! \cdot 5!}{(12-8)!}$$	8. How many different ways can the word PROBABILITY be arranged?

9. A study on the annual sales of a small business since its establishment in 2002 resulted in the following linear regression equation $y = 12x + 40$ where y is the annual sales (in thousands of dollars), and x is the number of years since 2002. Predict the annual sales in 2026.

10. Find the mean and complete the table to find the standard deviation of 5, 7, 2, 6, 13, and 3.

Data Values	$X - \bar{X}$	$(X - \bar{X})^2$
5		
7		
2		
6		
13		
3		

10. FINAL ASSESSMENT

Solve and record your answers. Time yourself (40 mins) and honestly assess what you've learned. Good luck!

1. Solve. $\dfrac{4^3 - 8 + \sqrt{-65 + 9^2}}{20 - 5^2 \cdot 2}$	2. Solve. $3(q - 2) + 2 = 5q - 7 - 2q$
3. Solve for b. $V = \dfrac{1}{3}bh$	4. Factor completely. $3x^2 - 9x - 30$
5. Evaluate. $\sqrt[3]{8x^4y^3} \cdot \sqrt[3]{27x^{12}y^{10}}$	6. Solve, graph, and write the solution in interval notation. $6 - \dfrac{2}{3}x < x - 9$

7. Luca has a collection of 15 dimes and quarters worth $2.40. How many of each coin does he have?

---

8. Complete below.

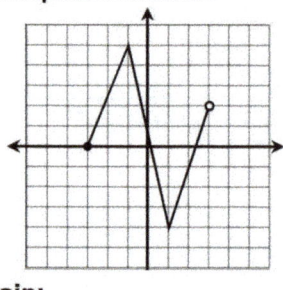

Domain:
_____

Range:
_____

Function?

9. Given $f(x) = -2x^2 + x - 5$ if f(x)=-20, find x.

---

10. Write in slope-intercept form, then graph. $2x + 5y = 10$

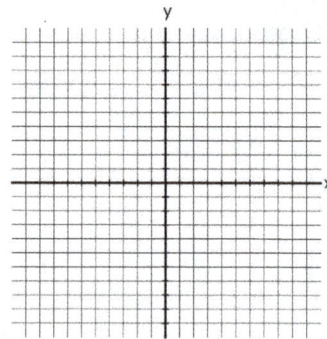

11. Write an equation of the line that passes through the points (0,2) and (4,6).

12. Write an equation of the line that passes through the given point and is (a) parallel to the given line and (b) perpendicular to the given line: $(4, -9)$, $y = \frac{1}{4}x + 2$

Parallel equation: _____

Perpendicular equation: _____

13. In an isosceles triangle, the length of each leg is three times the length of its base plus 2 meters. If the perimeter of the triangle is 60 meters, find the length of one of the legs.

14. Based on the system of equations, what is the value of 2x+3y?

$$y = 3x + 2$$
$$3x - 2y = -13$$

15. Multiply.

$$(2x + 1)(x - 1)$$

16. Solve.	17. If $f(x) = 3x^3 + 3x^2 + 9$
$6\|4x - 2\| - 2 = 16$	and $g(x) = 5x^3 - 7x^2 + 6x - 9$
	find *(f-g)(x)*.

18. Solve.	19. Solve.
$169x^2 - 1 = 0$	$4^{3x-30} = \left(\dfrac{1}{8}\right)^{3x}$

20. What is the solution set for the following system
of equations?
$$\begin{cases} -6x - y = -7 \\ y = 2x^2 - 10x - 41 \end{cases}$$

Answers:

IV. Prerequisite Assessment

1. 16
2. 5
3. 4
4. x=7
5. a=5
6. $\varnothing$
7. $(x+6)(x-6)$
8. {-4,-2}
9. -20
10. $\dfrac{5}{2}$ $x \neq -2$
11. x=3
12. $64x^{18}y^6$
13. $5x^2 + x + 3$
14. $x < -6$
15. see QR
16. (3,1)

SCAN ME

1. FA (Level 1)

1. x-5
2. 32
3. 10
4. -2
5. -7/4
6. y=-2x+3
7. 15
8. y=-2x+5
9. (5,0) (0,3)
10. see QR
11. y=-6x+3
12. $\mathbb{R}$
13. {-7,1}
14. 35 mi

SCAN ME

01. FA (Level 2)

1. 9(x+5)
2. -1/5
3. 47/7
4. f=4
5. g=4
6. $g = \dfrac{2+x}{3y}$
7. h=-21/11
8. $y = \dfrac{3}{4}x - \dfrac{15}{4}$
9. (2,0) (0,-5)
10. see QR
11. $y = 4x - 27$
12. $(-\infty, 1) \cup (1, \infty)$
13. {-3,13}
14. $\dfrac{29}{2}\pi$

SCAN ME

Answers:

01. FA (Level 3)

1. $x=12$
2. $w=-32$
3. $t=-24$
4. $x=-1$
5. $x=-\frac{17}{12}$
6. $w=\frac{3V}{hl}$
7. $x=-\frac{47}{13}$
8. $y=3x-7$
9. $(-10,0)$ $(0,6)$
10. see QR
11. perpen.
12. $[-4,4]$
13. $\{-8,16\}$
14. 20 hrs

02. Inequalities (Level 1)

1. $12x<6$
2. see QR
3. $x<-5.5$
4. see QR
5. $m\geq 3$
6. $x>7/3$
7. $x<24$
8. see QR
9. see QR
10. $x\geq 70$

02. Inequalities (Level 2)

1. $x+2\geq 9x$
2. see QR
3. $0\leq x<12$
4. see QR
5. $x<2$
6. $x<-13$
7. $\emptyset$
8. see QR
9. $-\frac{13}{2}\leq x\leq 3$
10. $h\geq 17.6\ hours$

Answers:

02. Inequalities (Level 3)

1. $\frac{x}{5} \geq 10$
2. see QR
3. $x < -7 \; or \; x > 1$
4. see QR
5. $n > -\frac{1}{9}$
6. $r < 4$
7. $\mathbb{R}$
8. see QR
9. see QR
10. $r \leq 21$

SCAN ME

03. Functions & Rel. (Level 1)

1. D: {2,-1,-5,0}
   R: {3,5,-7}
2. yes
3. {3, 21, -3}
4. h(x)=x+1
5. h(x)=21x+1
6. h(x)=x-6
7. x=4
8. {-4,7,9}
9. see QR
10. $x \to \infty,$ $y \to -\infty$ $x \to -\infty,$ $y \to \infty$

SCAN ME

03. Functions & Rel. (Level 2)

1. D: {-3,-2,0,2,3}
   R: {0,4,-3,5,1}
2. no
3. {3,0,-1}
4. h(x)=5x+3
5. h(x)=-8x+14
6. $h(x) = -\frac{3}{2}x + 10$
7. x=2
8. {2,-1,-14,0}
9. see QR
10. $x \to \infty,$ $f(x) \to \infty$ $x \to -\infty,$ $f(x) \to -\infty$

SCAN ME

Answers:

03. Functions & Rel.(Level 3)

1. $D: (-\infty, \infty)$
   $R: (-\infty, \infty)$
2. no
3. $\{1, 16\}$
4. $h(x) = 9x + 1$
5. $h(x) = -3x - 3$
6. $h(x) = 2x - 8$
7. $x = 16$
8. $\{3, 0.4, 25, 100\}$
9. see QR
10. $x \to \infty,$
    $f(x) \to \infty$
    $x \to -\infty,$
    $f(x) \to \infty$

SCAN ME

04. Sys. of Equations (L1)

1. $(1, -1)$
2. $(1, 0)$
3. $(-\frac{13}{9}, -10)$
4. $(-\frac{13}{3}, -3)$
5. see QR
6. see QR
7. 21 general admin, 14 early admin
8. 3 hardcover, 4 paperback

SCAN ME

04. Sys. of Equations (L2)

1. No solution
2. $\mathbb{R}$
3. $(-3, 5)$
4. $(\frac{67}{19}, \frac{7}{19})$
5. see QR
6. see QR
7. $78°, 102°$
8. 21 miles

SCAN ME

Answers:

04. Sys. of Equations (L3)

1. infinitely many
2. $(2,1)$
3. $(4,-3)$
4. infinitely many
5. see QR
6. see QR
7. 8 quarters, 16 dimes
8. 9 liters of orange, 6 liters of apple

SCAN ME

05. Quad. Funct. & Eq. (L1)

1. $3a^5 + 4a^2 - 7a$
   L.C:3
   D:5
2. $-2x^2 + 11x - 4$
3. h+5j
4. $12x^4 - 4x^3 + 10x^2$
5. $x^2 + 12x + 27$
6. $6x^3 - 16x^2 - 8x + 6$
7. $x^2 + 4x + 4$
8. $5(3x^2 + 20)$
9. $(x^2 - 3)(x + 2)$
10. $(x + 2)(x + 8)$
11. $(x + 5)^2$
12. $x = \pm 8$
13. 5
14. $x = \pm 2$
15. {-9,3}
16. {-8,4}

SCAN ME

05. Quad. Funct. & Eq. (L2)

1. $-2h^4 - 6h^3 + 5h - 9$
   L.C:-2
   D:4
2. $x^3 + 2x^2 + 5x - 1$
3. $-3x^2 - x + 7$
4. $-4x^3 + 13x^2 + 6x - 12$
5. $5w^2 + 13w - 6$
6. $b^3 - 6b^2 + 11b - 6$
7. $x^4 - 36$
8. $3x(2x^2 - 3x + 4)$
9. $(x + 1)(x^2 + 4)$
10. $(x - 2)(5x - 3)$
11. $(x + 10)(x - 10)$
12. x=5
13. $6\sqrt{2}$
14. $\pm\frac{5}{2}$
15. {-1,9}
16. {-1,3}

SCAN ME

87

Answers:

05. Quad. Funct. & Eq. (L3)

1. $13x^2 - 14x - 5$
   L.C: 13
   D: 2
2. $8x^2 - 3x - 1$
3. $2x^2 - 2x + 25$
4. $8x^3 - 18x^2 + x - 15$
5. $12k^2 - 31k + 7$
6. $2x^3 + 5x^2 - 4$
7. $\frac{1}{16}x^2 + x + 4$
8. $7xy(2x - 1)(x + 1)$
9. $(x + 6)(y - 6)$
10. $-(x + 1)(2x + 3)$
11. Prime
12. $\{0,4,7\}$
13. $4\sqrt{5}$
14. $\{-1,3\}$
15. $x = 3 \pm \sqrt{17}$
16. 37

SCAN ME

06. Quad. F. & Graphs (L1)

1. see QR
2. see QR
3. AOS, x=3
   V:(3,-4);
   y-int: (0,5)
4. $y = (x + 1)^2 - 4$
5. V:(-2,-5),
   AOS x=-2,
   y-int (0,-1), Range: $[-5,\infty)$
6. $y = (x + 1)^2 - 3$
7. (1,3) &(-1,7)
8. t=5 seconds

SCAN ME

06. Quad. F. & Graphs (L2)

1. see QR
2. see QR
3. AOS, x=-2
   V:(-2,-1);
   y-int: (0,7)
4. $y = (x + 6)^2 - 32$
5. V:(1,4),
   AOS x=1,
   y-int (0,3), Range: $(-\infty, 4]$
6. $y = -(x - 1)^2 + 3$
7. (5,-5) &
   (-1,-5)
8. 19 ft, 1 sec.,
   2.09 sec.

SCAN ME

Answers:

06. Quad. F. & Graphs (L3)

1. see QR
2. see QR
3. AOS, x=1
   V:(1,0)
   y-int: (0,-2)
4. $f(x) = -(x - 3)^2 + 11$
5. V:(-1,3),
   AOS x=-1,
   y-int (0,5), Range: $[3, \infty)$
6. $f(x) = 3(x - 4)^2 - 2$
7. No solution
8. see QR

SCAN ME

07. Exponents & Rad. (L1)

1. 998
2. 1/27
3. 1/4
4. $\frac{8}{x^4}$
5. 6
6. $x^{\frac{1}{4}}$
7. $\sqrt[7]{x^3}$
8. 4
9. 2
10. xy^3
11. $2\sqrt{5}$
12. $x^2\sqrt{x}$
13. y^9
14. $x^{\frac{1}{2}}$
15. 1/4
16. $12xy\sqrt{2y}$
17. $f(x) = \frac{1}{4}x - \frac{1}{4}$

SCAN ME

07. Exponents & Rad. (L2)

1. 12/49
2. 1
3. -2/3
4. $-5k^2$
5. 2
6. $a^{\frac{5}{2}}$
7. $3\sqrt{x^3}$
8. 1/3
9. 3
10. a^3b^5
11. $2\sqrt[3]{4}$
12. $x^5\sqrt[3]{x^2}$
13. $y^{\frac{3}{10}}$
14. $\sqrt[4]{11^3}$
15. $6x^{\frac{8}{15}}$
16. 1
17. $4x^2 - 6$

SCAN ME

Answers:

07. Exponents & Rad. (L3)

1. 8/9
2. -1/125
3. -9/8
4. $\frac{c^6}{a^2 d^3}$
5. 10
6. $(2x)^{\frac{7}{3}}$
7. $\sqrt[5]{(4a)^3}$
8. 125
9. 5
10. $3y^4 z^4 \sqrt[3]{z^2}$
11. $2\sqrt[4]{3}$
12. $3c^2 d \sqrt[4]{4d^2}$
13. $\frac{5y^6}{6x^3}$
14. 3/5
15. $\frac{-6}{x^4}$
16. $\frac{\sqrt{6}}{3}$
17. 3

SCAN ME

08. Exponential Funct. (L1)

1. 13,16, 19
2. expon.
3. growth, (0,1/3)
4. 9617
5. 16
6. $5634.13
7. 3
8. 1
9. -1/5
10. 7

SCAN ME

08. Exponential Funct. (L2)

1. 64,-128, 256
2. linear
3. growth, (0,0.455)
4. 304.78
5. 12.5
6. $3745.92
7. 4
8. 5
9. 1
10. 0

SCAN ME

Answers:

08. Exponential Funct. (L3)

1. $a_n = 9n - 11$ 6. 4%
   61   7. 9
2. exponential 8. 9/2
3. decay,   9. {-2,6}
   (0,3)   10. -9/5
4. $655.36
5. $1689.74

SCAN ME

09. Stats & Probability (L1)

1. 11.6   6. 16
2. 9   7. 13, 100,
3. 4, 8   22%
4. 50%   8. see QR
5. 8.3%   9. see QR
           10. 26.79%

SCAN ME

09. Stats & Probability (L2)

1. see QR   6. see QR
2. 32.5, 32, 7. 68%
   28, 32   8. 95%
3. 25, 15   9. see QR
4. 30 ways 10. 60
5. 28.16, 5.31

SCAN ME

Answers:

08. Exponential Funct. (L3)

1. $a_n = 9n - 11$
   61
2. exponential
3. decay,
   (0,3)
4. $655.36
5. $1689.74
6. 4%
7. 9
8. 9/2
9. {-2,6}
10. -9/5

SCAN ME

09. Stats & Probability (L1)

1. 11.6
2. 9
3. 4, 8
4. 50%
5. 8.3%
6. 16
7. 13, 100, 22%
8. see QR
9. see QR
10. 26.79%

SCAN ME

09. Stats & Probability (L2)

1. see QR
2. 32.5, 32, 28, 32
3. 25, 15
4. 30 ways
5. 28.16, 5.31
6. see QR
7. 68%
8. 95%
9. see QR
10. 60

SCAN ME

Answers:

09. Stats & Probability (L3)

1. 11, 9, N/A, 18
2. 41.7%
3. 9.1%
4. see QR
5. 1
6. see QR
7. 3600
8. 9,979,200
9. $328,000
10. 3.56

SCAN ME

10. Final Assessment

1. -2
2. No Solution
3. 3V/h
4. 3(x-5)(x+2)
5. $6x^5y^4\sqrt[3]{xy}$
6. x>9, see QR
7. (6,9)
8. [-3,3), [-4,5] Yes
9. x=-5/2, 3
10. y=-2/5x+2
11. y=x+2
12. $y = \frac{1}{4}x - 10$
   $y = -4x + 7$
13. x=8
14. 39
15. $2x^2 - x - 1$
16. $\left\{-\frac{1}{4}, \frac{5}{4}\right\}$
17. $-2x^3 + 10x^2 - 6x + 18$
18. $\left\{-\frac{1}{13}, \frac{1}{13}\right\}$
19. x=4
20. (6,-29) (-4,32)

SCAN ME

If you enjoyed this book, please consider leaving a review or follow me on instagram @thattck90 for future releases.

"Defeat is only a state of mind, and nothing more."
David J. Schwartz

ABOUT
JULIA WITCHARD

My journey into the world of education was as unexpected as it was transformative. As an unemployed graduate student, uncertain of my career path, studying international government and politics for the Foreign Service Officer Test (FSOT), my career path took an unforeseen turn when the principal of a high school approached me with an intriguing proposition – to explore a career in teaching.

Over the course of five years, I found myself deeply attached to my students, their challenges, and their triumphs. I devoted countless hours before and after school striving to make mathematics an enjoyable and relatable subject. My classroom became a hub of activities, a place where each student felt seen and heard. My students often questioned why I chose teaching, as if they could perceive a potential in me that I was still uncovering. Driven by a firm belief that every student matters, I refused to accept the notion that "you cannot save them all." This belief fueled my passion and led to the creation of engaging activities, initially shared on Teachers Pay Teachers, and now an interactive workbook. My aim has always been clear: to make a difference, one student at a time. Now, as I venture onto Amazon and other platforms, I see this not just as an expansion but as a step closer to "saving" them all. This is more than just a career for me; it's a mission to ignite a love for learning and a belief in the power of education.